ASTRONOMY ONLINE

30 interactive Web modules
for any astronomy course
www.whfreeman.com/aol

Timothy F. Slater
University of Arizona

W. H. Freeman and Company
New York

W. H. Freeman & Company

Media Editor   Charlie Van Wagner
Acquisitions Editor   Patrick Farace
Publisher   Michelle Julet
Marketing Manager   Jeffrey Rucker
Project Editor   Jodi Isman
Development Editor   Donald Gecewicz
Cover Design   Trina Shackleford

Programming, animation, course design, and development:
Sumanas, Inc.
Suite 390
131 N. El Molino Ave.
Pasadena, CA 91101
www.sumanasinc.com

ISBN: 0-7167-9669-4

© 2002 by W. H. Freeman and Company

No part of this book may be reproduced by any mechanical, photographic, or electronic process, or in the form of a phonographic recording, nor may it be stored in a retrieval system, transmitted, or otherwise copied for public or private use, without the written permission from the publisher.

Printed in the United States of America

First printing 2002

CONTENTS

Introduction *1*
What is *Astronomy Online*? *1*
What are the features of *Astronomy Online*? *2*
How can *Astronomy Online* be used in an astronomy course? *3*
How can *Astronomy Online* be accessed? *3*
Description of Modules *5*

MODULE 1	Astronomy and the Universe	*5*
MODULE 2	Knowing the Heavens	*6*
MODULE 3	Eclipses and the Motion of the Moon	*7*
MODULE 4	Gravitation and the Waltz of the Planets	*8*
MODULE 5	The Nature of Light	*9*
MODULE 6	Optics and Telescopes	*10*
MODULE 7	Our Solar System	*11*
MODULE 8	Our Living Earth	*12*
MODULE 9	Our Barren Moon	*13*
MODULE 10	Sun-Scorched Mercury	*14*
MODULE 11	Cloud-Covered Venus	*15*
MODULE 12	Red Planet Mars	*16*
MODULE 13	Jupiter: Lord of the Planets	*17*
MODULE 14	The Galilean Satellites of Jupiter	*18*
MODULE 15	The Spectacular Saturnian System	*19*
MODULE 16	The Outer Worlds	*20*
MODULE 17	Vagabonds of the Solar System	*21*
MODULE 18	Our Star, the Sun	*22*
MODULE 19	The Nature of the Stars	*23*
MODULE 20	The Birth of Stars	*24*
MODULE 21	Stellar Evolution: After the Main Sequence	*25*
MODULE 22	Stellar Evolution: The Deaths of Stars	*26*
MODULE 23	Neutron Stars	*27*
MODULE 24	Black Holes	*28*
MODULE 25	Our Galaxy	*29*
MODULE 26	Galaxies	*30*
MODULE 27	Quasars, Active Galaxies, and Gamma-Ray Bursters	*31*
MODULE 28	Cosmology: The Creation and Fate of the Universe	*32*
MODULE 29	Exploring the Early Universe	*33*
MODULE 30	The Search for Extraterrestrial Life	*34*

ACKNOWLEDGMENTS

All media produced by Sumanas, Inc.

Special thanks to Janelle Bailey for composing the online end-of-module exams. She helped out considerably and significantly added quality and value to *Astronomy Online*. Don Gecewicz, Development Editor for *Astronomy Online*, supported the actual "writing" part of the project every step of the way. His suggestions were always right on and out of this world. Charlie Van Wagner, Media Editor for *Astronomy Online*, first encouraged me to create this exciting new project. I thank him for taking the moving deadlines in stride. And thanks to Roger Freedman and Neil Comins, authors of the Freeman Astronomy texts, for allowing me to use the media assets originally created to supplement their textbooks. In the end, all errors are my responsibility.

T. F. Slater, Tucson, Arizona

I. INTRODUCTION

Over the last five years, personal computers and the Internet have given students the opportunity to explore actual scientific data sets and use scientific visualization tools in ways that only professional scientists could only a few years before. Initially, astronomy textbook authors were overjoyed with this multimedia technology because it allowed them to include CD-ROMs in textbooks that included many more images than could be fit into the text. In some cases, extensive graphic design animation allowed some previously static textbook diagrams to come alive on a computer screen. This tool also provided students with a plethora of movies that they could play over and over again at home while studying for exams. Unfortunately, the majority of the authors' dreams of fully engaging, animated textbooks were never realized because faculty did not have a highly structured way of assigning and monitoring how students were using information-filled CD-ROMs. With the quickly rising prices of textbooks, both faculty and students were left bewildered at how all of the effort and financial resources put into expertly building CD-ROMs had an insignificant impact on achievement.

During this same time, astronomers and science educators learned an enormous amount about how people learn. The most influential of these results is that people learn most effectively when they engage in a subject by committing to their initial ideas, critically testing new concepts against existing ideas, and actively monitoring their progress in learning new concepts. Expert learners with considerable practice in learning new subjects do these things when they read textbooks and listen to lectures. However, these are not the actions of most people.

As a direct result, I became interested and intrigued with the idea of using media to facilitate the education of astronomy students and began working on a new project called *Astronomy Online*.

II. WHAT IS *ASTRONOMY ONLINE*?

Going far beyond the idea of an animated textbook, *Astronomy Online* is a hypermedia product that uniquely weaves engaging multimedia visualizations into a structured learning environment that reflects what we now know about how people learn. The goal was to create a conceptual survey of introductory astronomy by dividing complex concepts into cognitively appetizing bite-size pieces. Each cognitive piece is clearly summarized with hyperlinks explaining all terms as well as illustrated using high-resolution images, insightful animations, and video presentations that can be accessed by any computer connected to the Internet.

> *Astronomy Online* is a hypermedia product that uniquely weaves engaging multimedia visualizations into a structured learning environment that reflects what we now know about how people learn.

Unique to *Astronomy Online*, each module helps students engage in the pursuit of astronomy by providing activities where students can make authentic astronomical observations using many of the same Internet data resources research scientists use. Further taking advantage of a hypermedia environment, *Astronomy Online* provides hyperlinks to in-depth information, images with comprehensive figure captions, videos, animations, living diagrams, and interactive exercises that provide multiple approaches to explaining complex concepts. These hyperlinks should ALWAYS be accessed and carefully studied by the student. In order to successfully complete the online examination found at the end of every module, students will need to access and carefully study every aspect of the media.

Consistent with how people learn, each instructional module begins by asking the student to answer questions about the concept under study to make their knowledge explicit. At the conclusion of each module, students are prompted to reconsider their initial answers and provide scientifically accurate explanations. Finally, the student is quizzed by a 20-question exam found at the end of every module. In total, *Astronomy Online* is a comprehensive online learning environment that is consistent with how people learn.

III. WHAT ARE THE FEATURES OF *ASTRONOMY ONLINE*?

Astronomy Online uniquely takes advantage of the hypermedia learning environment to engage students in the study of astronomy using a combination of a structured set of guided learning activities and illustrated explanations. The materials are carefully constructed upon a foundation of what researchers understand about how people learn, the common astronomy misconceptions students embrace, and some of the actual questions and Internet resources that astronomers use in their research.

The principle features of each module are:
- Overview of key questions or observations guiding the exploration of a concept
- Interactive activity that focuses on astronomical observations
- Pre-instructional questions to elicit students' misconceptions
- Carefully sequenced hypertext pages illustrated with animations and videos
- Regularly spaced concept-checks using drag-and-drop interactives
- Post-instructional questions comparing students' initial ideas to new ideas
- Interactive astronomy laboratory activities to increase students' reasoning skills
- Online, multilevel, computer-graded, multiple-choice exam that provides students with instant feedback on their progress
- E-mailed assignments from students to the professor, which significantly reduces the amount of hard copy paperwork associated with the course

In addition to providing students with targeted instruction in astronomy, *Astronomy Online* is designed to allow faculty to record and track how students are progressing through the material. Every module includes a 20-item, multiple-choice examination. The exams contain items that are equally divided among knowledge-level questions, comprehension-level questions, and advanced application-level questions. After students take the exam their answers can either be e-mailed directly to the professor's personal account or can be saved on an online database and retrieved at a later date.

IV. HOW CAN *ASTRONOMY ONLINE* BE USED IN AN ASTRONOMY COURSE?

Astronomy Online can be easily integrated into the introductory astronomy survey course in at least three distinct ways, and possibly more. Probably the most common way *Astronomy Online* will be used is as an illustrated and animated supplement to the traditional lecture course and textbook. Students should be assigned to complete both the activity and the interactive reading portions of the modules related to the course to fully benefit from the integrated nature of the modules. A second way to use the modules is as a computerized interactive tutor that engages students in reasoning exercises about fundamental concepts by requiring students to answer questions before proceeding to more advanced concepts. A third way faculty can use *Astronomy Online* is as a complete set of materials for students doing an independent study in astronomy. As the instructor, *Astronomy Online*'s computerized monitoring features allow you to keep track of all your students' progress easily so that any problems or insufficient effort with an independent study program can be recognized and corrected early in the semester.

V. HOW CAN *ASTRONOMY ONLINE* BE ACCESSED?

It's simple to access *Astronomy Online*.

- Go to the companion web site at www.whfreeman.com/aol/
- Click the link "Web Version" located under the text "Log on to Astronomy Online."
- Enter all required information (i.e., name, e-mail address, etc.).
- Enter the access code located on the inside front cover of this booklet. (NOTE: Once this code is used to create a unique password it will cease to function.)
- Click on the "Access Astronomy Online" link located at the bottom of the screen.
- Log in using your e-mail address and newly created password.

If you are an instructor, *Astronomy Online* has been designed to allow you to record and track how students are progressing through the material. As mentioned earlier, each module in *Astronomy Online* includes a 20-item, multiple-choice examination. The exams contain items that are equally divided among knowledge-level questions, comprehension-level questions, and advanced application-level questions. The results of these quizzes are saved on an online database. You can access the answers by simply clicking the "For Instructors" link on the home page, entering an instructor password, and keying in your e-mail address. If you do not have an instructor password, you will be prompted to register for security reasons. Once we have confirmed that you are, in fact, an instructor we will e-mail you your password so you can access your students' results. Results can be stored on this database or you can have them e-mailed to your personal account.

If you encounter any problems with registration or any other technical difficulties, please contact our technical support department at:

    techsupport@bfwpub.com
    or
    1-800-936-6899

VI. DESCRIPTION OF MODULES

MODULE 1. Astronomy and the Universe

Objects in the Sky

In the space provided, list as many different things as you can see in the sky (daytime and nighttime). Correct spelling is not as important as getting down as many ideas as possible.

To save your answer, click the "Save Answer" button before you leave this page. When you are finished with this module, go to the **Module Conclusion** and follow the instructions for sending this information to your course instructor.

You are not required to provide the correct answer to get full credit. But you must give a detailed and thoughtful response, which will be shown to you again at the end of this tutorial. When you send your response to the course instructor, you will receive credit for a thoughtful response.

Tutorial: Objects in the Sky, page 1 of 5

Staring up into a star-filled night, many of us wonder about the universe hovering over our heads. Some of our questions are philosophical, such as, "Why are we here?" Others are more scientific, "Exactly how far away are those stellar points of light?" Today, astronomy is the scientific study of the cosmos. Astronomers use the latest technology to decode hidden information in starlight, powerful computers to model and simulate the interior of the Sun, and robotic space probes to explore the nearby planets of our solar system. Although our knowledge is rapidly growing, there are still many mysteries to be solved.

MODULE 2. Knowing the Heavens

The Northern Sky

Constellations are semirectangular regions of the sky. Within those regions, astronomers often draw connect-the-dots shapes to make the constellations easier to identify. These connect-the-dots patterns of familiar shapes are known as **asterisms**. Asterisms sometimes fall within a constellation, such as the Big Dipper in Ursa Major. Ursa Major represents the Big Bear but the seven brightest stars within that constellation make a ladle shape.

(Click image to enlarge.)

Looking toward the northern sky on any clear night of the year, the most easily identifiable asterism is the Big Dipper. A line drawn between the two stars that make up the end of the dipper's cup and extended to the next bright star indicates the position of Polaris, the North Star. The North Star is the last star in the handle of the Little Dipper. Like the Big Dipper, the Little Dipper is composed of seven stars in the shape of a ladle. However, these stars are much dimmer, making this pattern much more difficult to see.

If you trace the bending arc of the Big Dipper's handle, you come to a very bright star in the constellation of Boötes called Arcturus. Boötes looks like a giant kite or an ice cream cone to many people. And if you continue to follow the arc of the Big Dipper's handle past Arcturus, you come to the brightest star in Virgo, named Spica. Memorizing patterns such as this is an easy way to begin to learn the stars in the night sky.

Helpful phrases for learning the night sky:
- Follow the arc to Arcturus and spike on to Spica.
- One of the dippers always looks as if it is pouring water into the other dipper.
- The last two stars in the cup of the Big Dipper always point to Polaris.
- The inner two stars in the cup of the Big Dipper always go right to Regulus.

Asterisms are highly personal things and can vary considerably from one star map to the next. Over time, you'll probably develop your own personal set of asterisms to help you find regions and objects in the sky.

Tutorial: Constellations and the Celestial Sphere, page 3 of 8

One of the most magical parts of astronomy involves going outside on a clear night and looking up into the star-filled sky. Careful observers find bright stars and dim stars, red stars and white stars, clusters of countless stars and areas with few stars, and regions with so many stars that part of the sky almost glows. It is not difficult to imagine that generations of earthlings marveled at the night sky even before the advent of written history. Today, astronomers recognize 88 different semirectangular constellations covering the entire sky. Many of the constellations carry names known to us from myths and legends of antiquity. The most familiar constellations include Orion the Hunter, Cygnus the Swan, Ursa Major the Giant Bear, and Canis Major the Great Dog. Most importantly, understanding changes in the positions and motions of the Sun, the Moon, and the stars has had a deep impact on our culture and still fundamentally underlies how astronomers measure the passage of time on daily and yearly scales.

MODULE 3. Eclipses and the Motion of the Moon

 Self Test: Phases of the Moon

Match the most likely name of each object to its image by dragging the (?) symbol next to each name to the (?) symbol for each picture.

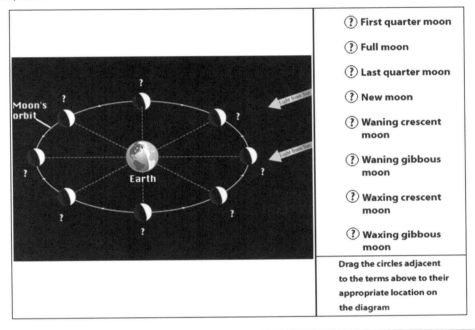

(?) First quarter moon

(?) Full moon

(?) Last quarter moon

(?) New moon

(?) Waning crescent moon

(?) Waning gibbous moon

(?) Waxing crescent moon

(?) Waxing gibbous moon

Drag the circles adjacent to the terms above to their appropriate location on the diagram

Tutorial: Motion of the Moon, page 6 of 9

Possibly the most frequently noticed night-sky object is our Moon. A solid sphere of rock orbiting our planet at a distance of 360 thousand kilometers, the Moon's cyclical change in appearance over the course of a month has been the basis of calendars and timekeeping for many cultures throughout history. The Moon holds a prominent place in the lives of people besides astronomers. It has influenced storytellers, poets, painters, and lovers. Even the movement of animals and ocean tides cannot escape the Moon's influence.

MODULE 4. Gravitation and the Motions of the Planets

Johannes Kepler

Although Brahe himself was not able to use his wealth of data effectively to come to strong conclusions about the workings of the solar system, he hired a young mathematician named Johannes Kepler (1571-1630), who developed three general principles that describe how planets move around the Sun. Now called Kepler's laws, these principles were based on Brahe's extensive and very accurate observations of planetary positions rather than any theory about why the planets should move as they do.

(Click image to enlarge.)

Astronomers had long assumed that all heavenly objects must move in perfectly round circles. In stark contrast, Kepler found that if he assumed that planets moved in noncircular orbits, called ellipses, he could more precisely represent the motions of the planets. This critical discovery resulted in **Kepler's first law**:

The planets orbit the Sun following the path of an ellipse with the Sun off-center at a focus point.

This statement was profound for two reasons. First, it was a definite statement about a moving Earth around a stationary Sun in a noncircular path. Second, the Sun was not at the center of the orbit but rather offset from the center at one focus.

Since the dawn of time, ancient astronomers watched the sky and tracked the motions of the heavens. Standing out among the thousands of twinkling stars, there were five "star-like" objects that captured early astronomers' attention. These objects seemed to wander among the constellations and often changed their brightness. Being able to predict the exact movements of these objects, which we now call planets, served as the seminal challenge for Copernicus, Brahe, Kepler, Galileo, and Newton, who dared to try to unlock the secrets of the solar system.

MODULE 5. The Nature of Light and Matter

Considering Star Colors and Temperatures

When looking up at the night sky, you might notice that stars appear in a variety of brightnesses and colors. Five bright stars and their temperatures are listed below. Estimate the color each would appear to be in the sky and explain your reasoning.

1. Aldebaran (approximate temperature 4000 K)
2. Pollux (approximate temperature 4200 K)
3. Sun (approximate temperature 5800 K)
4. Procyon (approximate temperature 6700 K)
5. Altair (approximate temperature 7000 K)

(Click image to enlarge.)

To save your answer, click the "Save Answer" button before you leave this page. When you are finished with this module, go to the **Module Conclusion** and follow the instructions for sending this information to your course instructor.

You are not required to provide the correct answer to get full credit. But you must give a detailed and thoughtful response, which will be shown to you again at the end of this tutorial. When you send your response to the course instructor, you will receive credit for a thoughtful response.

Tutorial: Light and Temperature, page 1 of 9

Astronomers are passionately interested in the nature of light because it is the only information that we get from stars. The stars, and most of the planets for that matter, are so far away that we will likely never visit them. So, for now, astronomers must learn to decode information hidden in starlight if we are to learn anything at all about the nature of these distant objects. Fortunately, the properties of light are reasonably well understood. After light is split into a rainbow by passing it through a prism or similar device, careful inspection of the light provides detailed information about the temperature, composition, and motion of even the most distant of stars.

MODULE 6. Optics and Telescopes

Telescope Magnification Exercise

A common telescope found in department stores is a 760-mm (3-inch) diameter refractor with a focal length of 600 mm that boasts magnification of 300 times. What magnifications are achieved by using each of the following commonly found eyepieces?

1. Eyepiece A with focal length 40 mm

To save your answer, click the "Save Answer" button before you leave this page. When you are finished with this module, go to the **Module Conclusion** and follow the instructions for sending this information to your course instructor.

2. Eyepiece B with focal length 25 mm

To save your answer, click the "Save Answer" button before you leave this page. When you are finished with this module, go to the **Module Conclusion** and follow the instructions for sending this information to your course instructor.

3. Eyepiece C with focal length 12 mm

To save your answer, click the "Save Answer" button before you leave this page. When you are finished with this module, go to the **Module Conclusion** and follow the instructions for sending this information to your course instructor.

4. Eyepiece D with focal length 2.5 mm

To save your answer, click the "Save Answer" button before you leave this page. When you are finished with this module, go to the **Module Conclusion** and follow the instructions for sending this information to your course instructor.

Telescopes serve as the astronomer's principal tool to gather light from distant objects so that it can be analyzed. Over the years, telescope designs have changed substantially. The first telescopes were made of two small glass lenses. Since those first simple (and limited) telescopes, the need for larger telescopes led telescope makers to use mirrors to collect light rather than heavy glass lenses. Today, the largest research telescopes use a combination of enormous mirrors and precise computer controls to collect every possible photon from a distant and dim object. These large modern telescopes are located on mountaintops above much of Earth's blurring atmosphere, or on satellites orbiting Earth. New technology is allowing astronomers to peer into parts of the universe never seen before.

To use this calculator, enter the focal lengths into the appropriate boxes, then click on the **OK** button to determine the magnification. Note: "e" stands for a power-of-ten exponent. For example, 1.0e3 = 1.0×10^3, 1e-3 = 1.0×10^{-3}.

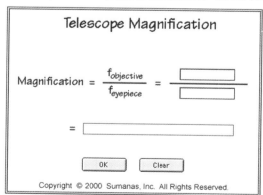

MODULE 7. Our Solar System

Formation of Stars and Planets

Stars form from the collapse of enormous clouds of dust and gas. In the process of collapsing, a slowly rotating cloud begins to spin faster and faster and becomes hotter. As shown in the video The Birth of the Solar System (more info), in collisions between the outer edges of the collapsing cloud, particles begin to combine together into larger particles in a process called **accretion** (more info). Over a few million years, dust and pebbles combined into roughly a billion asteroid-like objects called planetesimals with diameters of about 10 km. Over the next million years, these planetesimals collided, melted, and combined into still larger objects called protoplanets, which eventually coalesced into today's planets.

The planets formed by the accretion of planetesimals and the accumulation of gases left over from the dust and gas cloud that collapsed to form the Sun.

Astronomers have observed regions around infant stars in various stages of this process today. The first observable hint of stars surrounded by orbiting dust disks came from observations of a star called Beta Pictoris (more info). A more recent example observed from the Hubble Space Telescope is protoplanetary disks in the Orion Nebula (more info) called proplyds (more info).

Is this the correct account of the formation of the planets? If the matter in the initial cloud was all spinning in the same direction, it seems plausible that any planets resulting from such a process would also orbit the Sun in the same direction. As a secondary confirmation, this process also accounts for the distribution of chemical elements observed in the solar system. The temperature of the initial solar nebula would be warmer closer to the Sun and cooler farther away (more info). At the orbital distances of the terrestrial planets, the temperature would have ranged from roughly 500 to 1500 K. At these temperatures, the rock-forming elements would begin to condense. Farther from the Sun, the temperatures would have been much lower, perhaps 50 to 150 K, which is the condensation temperature for hydrogen and helium. This idea of varying temperatures causing different chemical elements to condense at different distances is consistent with observation of the formation of rocky, terrestrial planets near the Sun and gaseous, Jovian planets far from the Sun.

(Click any image on the right to enlarge.)

This sequence of drawings shows six stages in the formation of the solar system. (a) A slowly rotating cloud of interstellar gas and dust begins to contract because of its own gravity.

(b) A central condensation, the protosun, forms as the cloud flattens and rotates faster.

(c) A flattened disk of gas and dust surrounds the protosun, which has begun to shine.

(d) The Sun's rising temperature removes the gas from the inner regions, leaving dust and larger debris revolving in place.

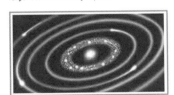

(e) The planets have established dominance in their regions of the solar system.

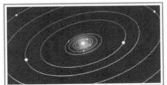

(f) The solar system as it appears today.

Aside from the Sun, the Moon, and the twinkling stars in the night sky, the planets are the next sky objects that sky watchers notice. Distinguished visually from stars because they do not twinkle, these seemingly star-like objects seem to wander slowly among the unchanging constellations. The five planets that can be seen without a telescope provide just a hint of the many wondrous objects that orbit our Sun. Until about 40 years ago, very little was known about these objects. Then space probes were successfully launched to explore the solar system. Today, we have enormous image libraries of many of the planets, some of which have come from robotic probes that we have landed on their surfaces. Yet despite a wealth of detailed pictures, there are still many things that we do not understand about our solar system.

MODULE 8. Our Living Earth

Motions Inside the Earth's Mantle Cause Plate Tectonics

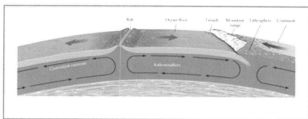

(Click image to enlarge.)

Because heat is trying to escape from Earth's outer liquid core, material in Earth's mantle is constantly expanding and moving upward toward Earth's crust. This process is called convection and is much like water boiling in a pot (more info) on the stove. The result of convection is the phenomenon known as plate tectonics.

The upwelling of material in the form of convection causes portions of Earth's surface, called plates, to move with respect to one another. Often, the collisions (more info) of these plates result in earthquakes. In some locations, such as down the middle of the Atlantic Ocean (more info), plates are moving apart in a process called seafloor spreading. As the plates separate (more info), molten material moves up, filling the space between the plates, resulting in an undersea mountain range, or oceanic ridge. In other locations, plates are colliding more info). If two plates are predominantly made of lightweight continents, the two plates buckle up, causing an uplifted mountain range such as the Himalayas (more info). If two colliding plates are composed of predominantly dense ocean floor material, then the two plates will create an island arc (more info).

Tutorial: Earth's Interior Shapes the Surface, page 4 of 7

Our home planet, Earth, is often called the blue planet. Nearly three-fourths of Earth's surface is covered in water, making it unique in the solar system. The remaining one-fourth of Earth's surface is land with highly varying terrain. Mountain ranges reach toward the sky; below, wide plains carved by rivers extend to the seas. Although the mountains and the plains seem unchanging, they are, in fact, changing slowly and significantly over time. First, forces emanating from deep within the Earth are altering many of the mountains. Second, Earth's weather significantly changes the landscape. Earth's weather is caused by rapid changes in our atmosphere, which is composed mainly of nitrogen with some oxygen. The oceans and the atmosphere absorb much of the Sun's energy and quickly transfer this energy back and forth in the form of wind, waves, and weather systems that alter the landscape. If scientists can understand the forces and phenomena that shape our Earth, this understanding will serve as a foundation for studying the other planets in our solar system.

MODULE 9. Our Barren Moon

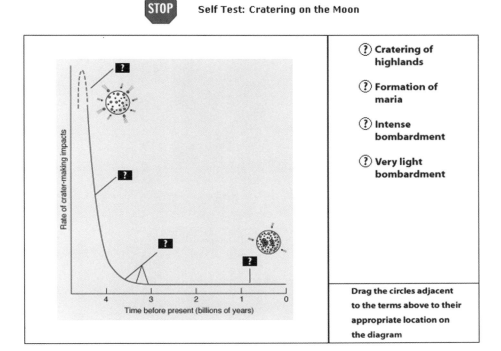

Tutorial: Our Barren Moon, page 6 of 9

The Moon is clearly one of the most dramatic sights in the nighttime sky. Visible about one half of each month in the nighttime and about half of each month in the daytime, the Moon is our planet's closest neighbor, at about 340,000 kilometers or 240,000 miles from Earth. Because the Moon has no weather or currently active lava flows, its ancient surface still holds many clues about the environment of the early solar system. When astronauts visited the Moon between 1969 and 1972, they left scientific instruments and brought back 4.6-billion-year-old rock samples that have provided scientists with enormous insight into the history of the cratered surface of the Moon and our solar system.

MODULE 10. Sun-Scorched Mercury

Observation: Sun-Scorched Mercury

(Click image to enlarge.)

1. In the figure, the two objects are shown at the same scale. Using a ruler on the screen, determine how many times larger than the Moon Mercury is. (*Hint:* Divide the measured diameter (top-to-bottom) of Mercury in millimeters by the measured diameter (top-to-bottom) of the Moon in millimeters).

2. Besides diameter, what is the most obvious difference in the surfaces of the two objects?

3. Compare the sizes of craters on the two objects. Excluding the large lunar maria, has one object apparently been hit with larger asteroids than the other? Why or why not?

The planet nearest to the Sun, Mercury, is so close to the brilliant Sun that it is difficult to study. Even compared to Pluto, astronomers probably know the least about Mercury despite several space probe flybys by *Mariner 10* in 1974. Mercury does have a crater-covered, lunar-like surface. However, Mercury does not have the characteristic dark-colored lava flows that we observe on the Moon. In 2008, a new probe called MESSENGER will arrive at Mercury with the intention of helping astronomers understand the exact character of the surface and the interior of this tiny planet.

MODULE 11. Cloud-Covered Venus

Cloud-Covered Venus Activity

Access the animation "Magellan Maps a Planet". If *Magellan* maps a strip about 25 km wide, what would be the minimum number of orbits *Magellan* would have to make to cover the entire circumference of Venus (Venus has a diameter of 12,104 km)? Show your work.

To save your answer, click the "Save Answer" button beside each question below. When you are finished with this module, go to the **Module Conclusion** and follow the instructions for sending this information to your course instructor.

(Click image to enlarge.)

The large circular structure near the center of this image is Aine Corona, a shallow dome some 200 km in diameter pushed upward by rising magma from Venus's mantle. The corona is surrounded by a number of smaller volcanic domes, called "pancake domes" for their shape, as well as by prominent and complex fracture patterns. Coronae (from the Latin for "crown") are unique to Venus; these geologic features have not been found on any other planet. Aine Corona lies in the vast plain to the south of Aphrodite Terra. (NASA)

Activity: Cloud-Covered Venus, page 1 of 1

Shining brilliantly first in the evening and then several months later in the morning, Venus is often the brightest object in the sky after the Sun and the Moon. Much of Venus's brilliance comes from highly reflective clouds that perpetually cover the planet. These obscuring clouds, however, have made it extremely difficult to study the surface. Only recently have astronomers perfected radio techniques to map the hidden surface precisely. Nearly identical in size to Earth, Venus, as we now know, is a planet of enormously high temperatures, acidic clouds, and odd surface geology that may show aspects of both ancient and newly resurfaced terrain.

MODULE 12. Red Planet Mars

Observation: Red Planet Mars

Observing Current Conditions on Mars

Access NASA's Internet site *Mars Today* and list the current wind speed, wind direction, temperature, and atmospheric water column at a latitude similar to where you are on Earth at this moment.

To save your answer, click the "Save Answer" button beside each question below. When you are finished with this module, go to the **Module Conclusion** and follow the instructions for sending this information to your course instructor.

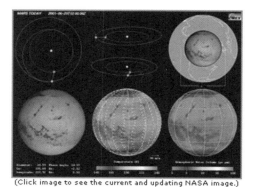

(Click image to see the current and updating NASA image.)

Observation: Red Planet Mars, page 1 of 1

As the fourth planet from the Sun, planet Mars probably holds the position as the most famous planet because of the imagination it has engendered in science fiction writers as a place life might exist beyond Earth. In the end, science fiction writers may be right. Mars is one of the most likely places where we may yet find evidence for extinct or even currently thriving life on another planet. Also, Mars is close enough that we might successfully land astronauts there in your lifetime. Mars has a thin carbon-dioxide atmosphere and experiences distinct seasons evidenced by rapidly growing and shrinking polar ice caps. Its surface is covered with enormous but extinct volcanoes. Most recently, orbiting space probes have uncovered clear evidence that water once flowed on its surface in years past.

MODULE 13. Jupiter: Lord of the Planets

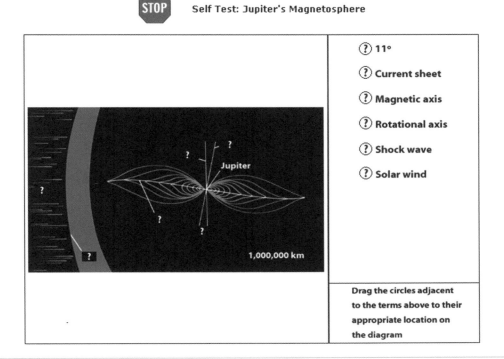

Tutorial: Jupiter: Lord of the Planets, page 9 of 10

Planet Jupiter is the largest of the objects orbiting our Sun. Orbiting at a distance more than five times greater than the size of Earth's orbit, Jupiter is a giant sphere with a mass more than 318 times that of Earth and a diameter more than eleven times that of Earth. Fundamentally different from the four terrestrial planets, Jupiter is primarily composed of hydrogen and helium gas with no solid surface. Its "atmosphere" is a dynamic and churning system that distinguishes itself from other planets by having easily observable, intricate light- and dark-colored bands and a swirling vortex known as the Great Red Spot. Jupiter is a world indeed worthy of the title *Lord of the Planets*.

MODULE 14. The Galilean Satellites of Jupiter

Planning Observations of Jupiter

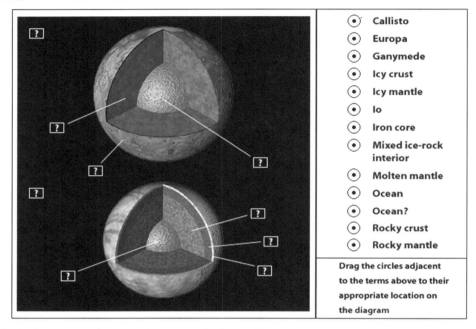

Describe the features that allow you to determine which of the Galilean satellites is which. You are required to provide a detailed and thoughtful answer to get full credit. When you send your response to the course instructor, and you will receive credit for being thoughtful and accurate.

To save your answer, click the "Save Answer" button beside each question below. When you are finished with this module, go to the **Module Conclusion** and follow the instructions for sending this information to your course instructor.

Activity: The Galilean Satellites of Jupiter, page 1 of 1

In the past, astronomers did not consider natural satellites, or "moons," orbiting other planets to harbor many interesting secrets. Indeed, compared to Earth's dynamically changing atmosphere, explosive volcanoes, and vibrant oceans, Earth's Moon is quiet and unchanging. Our views of how dynamic and exciting planetary moons can be were drastically changed in 1979 when the *Voyager I* and *Voyager II* missions flew past Jupiter and sent back high-resolution images of its four largest moons. Studied in greater detail by the *Galileo* mission in 2000, these four largest moons of Jupiter, called the Galilean Moons after their discoverer, Galileo Galilei, who first sighted them in 1610, form an exciting miniature planetary system surrounding Jupiter that may even be the most likely place to find life flourishing beyond Earth.

MODULE 15. The Spectacular Saturnian System

The Disappearing Ring Trick

Although Saturn has been known since ancient times, its ring system was discovered fairly recently. When Galileo first observed Saturn close up in 1610 with his small telescope, he didn't know what to make of two unresolved and puzzling lumps on either side of tiny Saturn. To make matters even more perplexing, these unexplained protrusions around Saturn seemingly disappeared in 1612 only to return in 1613. Forty years later and using a much better telescope, Dutch astronomer Christiaan Huygens discovered that a thin, flattened ring (more info) surrounding Saturn seems to appear and disappear at different times in Saturn's orbit.

Although Saturn's rings have a diameter of more than 274,000 km, they are only a few kilometers thick and are tilted at a constant angle of 27f to its orbital plane. As we observe Saturn from Earth slowly moving around the Sun, we view the ring system from various angles. At certain times, Saturn's north pole is tilted toward the Earth and Earthbound observers look down on the top side of the rings. At other times, our line of sight is directly in line with the Saturn's equatorial ring plane and the ultra-thin rings seem to disappear magically for several months. This strange phenomenon occurs about every 15 years and will occur again in 2008-2009, making Saturn's rings hard to observe from Earth.

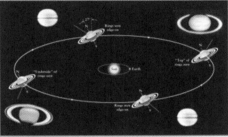

(Click image to enlarge.)

Saturn's rings are aligned with its equator, which is tilted 27° from the plane of Saturn's orbit. Therefore, Earth-based observers see the rings at various angles as Saturn moves around its orbit. Note that the plane of Saturn's rings and equator keep the same orientation in space as the planet goes around its orbit, just as the Earth does as it orbits the Sun. The accompanying Earth-based photographs show Saturn at various points in its orbit. Note that the rings seem to disappear entirely when viewed edge-on, which occurs about every 15 years. (Lowell Observatory)

Probably the most widely known planet is magnificently ringed Saturn. Nearly ten times larger than Earth, Saturn is almost as big as Jupiter. Revolving twice as far from the Sun as Jupiter, this gas giant is not only surrounded by countless ice chunks that comprise a brilliantly reflective ring system, but more than 20 moons also orbit Saturn. Saturn's largest moon, Titan, has a thick atmosphere of nitrogen and carbon dioxide, possibly with enough density to experience hydrocarbon rain. Saturn is indeed a spectacular sight.

MODULE 16. The Outer Worlds

Displaced Planetary Magnetic Fields

Voyager 2's magnetometer provided unexpected results of Uranus's and Neptune's magnetic fields. The magnetic fields of Earth, Jupiter, and Saturn are all closely aligned with their rotation axes. In other words, a compass on Earth points fairly close to the northern rotation axis. Surprisingly, on Uranus and Neptune, the planetary magnetic poles are offset by 59° and 47° from their rotation axes. Adding to the mystery, the magnetic fields of Uranus and Neptune are offset from the centers of the planets. At present, astronomers are at a loss as to how to explain these odd phenomena.

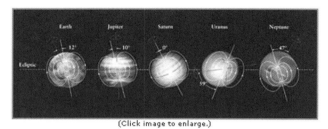

(Click image to enlarge.)

This drawing shows how the magnetic fields of Earth, Jupiter, Saturn, Uranus, and Neptune are tilted relative to their rotation axes. Note that the magnetic fields of all four Jovian planets are oriented opposite to that of Earth; on a Jovian planet, the north pole of a compass needle would point southward, not northward. Note also that the magnetic fields of Uranus and Neptune are offset from the center of the planets and steeply inclined to their rotation axes.

Tutorial: The Outer Worlds, page 8 of 12

No one knows who discovered the five wandering planets—Mercury through Saturn—because they have been observed and tracked since the dawn of civilization. However, the outer planets (Uranus, Neptune, and Pluto) were far too dim to be recognized as planets before the invention of large telescopes. One planet each was discovered in the eighteenth, nineteenth, and twentieth centuries. These planets were simply featureless dots in the heavens until the last 20 years when each has been closely monitored by the Hubble Space Telescope. Two of the three were also visited by the *Voyager 2* interplanetary space probe. We now know that these worlds have dynamically changing appearances and curious moons. Uranus and Neptune possess intricate ring systems and odd magnetic fields. Investigations of these worlds have generated more questions than answers for astronomers.

MODULE 17. Vagabonds of the Solar System

Observation: Vagabonds of the Solar System

The Distance to Near-Earth Objects (NEOs)

Astronomers closely monitor more than 300 of the thousands of objects that pass close to the Earth. Access NASA's *Near Earth Object Program* online at URL: http://neo.jpl.nasa.gov/ and answer the following questions.

1. Determine the name of and distance to the object that is closest to the Earth today.

To save your answer, click the "Save Answer" button beside each question below. When you are finished with this module, go to the **Module Conclusion** and follow the instructions for sending this information to your course instructor.

2. Determine the name of and distance to the object that will be closest on your birthday.

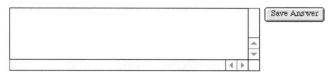

To save your answer, click the "Save Answer" button beside each question below. When you are finished with this module, go to the **Module Conclusion** and follow the instructions for sending this information to your course instructor.

3. Determine which object gets closest to the Earth and on which dates it occurs.

To save your answer, click the "Save Answer" button beside each question below. When you are finished with this module, go to the **Module Conclusion** and follow the instructions for sending this information to your course instructor.

4. Which of these objects are closer to the Earth than the Moon? Explain how you arrived at your answer.

To save your answer, click the "Save Answer" button beside each question below. When you are finished with this module, go to the **Module Conclusion** and follow the instructions for sending this information to your course instructor.

Observation: Vagabonds of the Solar System, page 1 of 1

Most of us already know that the solar system includes the Sun and the nine planets. Although these are the largest members of the solar system, they certainly are not the most numerous. Also orbiting the Sun relatively nearby are hundreds of thousands of rocky objects called asteroids and millions of icy objects—known as comets—that orbit much farther away. These objects are not often easily visible without a telescope, but occasionally comets appear close by, stretching brightly across the sky. Sometimes, these often hard-to-find objects do break into smaller pieces that can be easily seen. When these objects collide with Earth's atmosphere, they begin to glow. These phenomena are shooting stars, more formally known as meteors, and are visible on any clear night with a dark sky. These ancient objects hold critical clues for astronomers about the early formation of planetary systems.

MODULE 18. Our Star, the Sun

Energy Transport Inside the Sun

Once energy leaves the Sun's core, it takes a very long time for the energy, in the form of gamma rays released in the proton-proton chain, to make its way 696,000 km out to the photosphere. As gamma-ray photons are released, they do not travel very far before colliding with a nearby ion. When they bounce off that ion, they may or may not move in a direction pointing outward before they collide with yet another ion. The result is that it can take nearly a million years for photons to make their way out of the Sun.

Because the density is highest in the inner half of the Sun, collisions among ions and photons increase the temperature. This energy escapes the core through the process of radiative transfer by transferring energy to the more distant and cooler regions. At larger distances from the core, the Sun's density is considerably less, and a different energy transfer process takes place. Hotter regions expand and then float towards the photosphere because they are less dense. The overlying cooler regions then sink towards the center, and the process repeats itself. This churning motion where hot material moves upward and cooler material sinks is called convection. The result is that there are three distinct regions to the Sun's interior: the core where thermonuclear fusion is taking place, the radiative zone where energy is transferred via radiative transfer, and the outer convective envelope where material is actually moving up and down in circular cells.

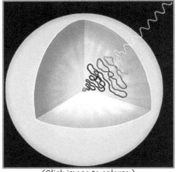
(Click image to enlarge.)

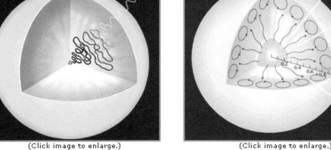
(Click image to enlarge.)

Tutorial: Solar Interior, page 5 of 9

Most of us think of the Sun as a quiet ball of steadily burning gas that provides the light and heat necessary for life to exist on Earth. As it turns out, our Sun is anything but quiet. Unpredictable explosions on the surface release tons of material that race out into the solar system causing communication satellites to malfunction, power grids on Earth's surface to overload, and the eerie northern and southern lights to glow at high latitudes. Inside the Sun, nuclear fusion operating at 15 million degrees combines hydrogen atoms into larger helium atoms in a process that releases energy that keeps the Sun from collapsing under its own weight.

As it turns out, many of the stars in the night sky are very similar to our Sun. If astronomers can understand the physical mechanisms that cause our Sun to act as it does, then those same principles should apply to the many distant Suns we see in the heavens. Understanding the physical processes that govern our Sun will help us better understand the distant stars twinkling in the night sky.

MODULE 19. The Nature of the Stars

Star Sizes

The stars are so far away that even in our largest telescopes the stars appear only as tiny points of light. Pictures of the stars (more info) show stars slightly smeared out into a disk, but this is due to the limitations of the film or imaging technique and provides no insight into the actual size of the stars. Astronomers determine the size of a star by combining information about its luminosity and its spectral class. The relationship of a star's luminosity, radius, and spectral class/temperature is defined by the Stefan-Boltzmann law. According to the Stefan-Boltzmann law, hot stars are generally bright and cool stars are generally dim. Further, if a star is cool, it can be bright only if it is very large. Likewise, if a star is hot, it will be quite bright unless it is quite small.

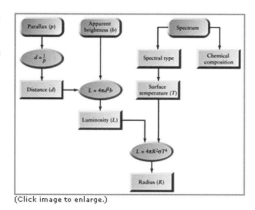
(Click image to enlarge.)

Stefan-Boltzmann law

$L = 4\sigma R^2 \sigma T^4$
L = star's luminosity in watts
R = star's radius in meters
σ = Stefan-Boltzmann constant = 5.67×10^{-8} W m^{-2} K^{-4}
T = star's surface temperature in kelvins
[mathematical details of the equation] (more info)

Tutorial: Stellar Spectra, page 4 of 9

A long look up into the night sky reveals that all stars are not created equal. Of the 3000 or so stars visible to the unaided eye each night, we can see that some are bright while most are dim. A pair of binoculars or a small telescope shows that there are thousands of stars dimmer than what can be seen without a telescope. In fact, astronomers estimate that there are more than 100 billion stars in our Milky Way Galaxy alone.

By looking closely at stars in the night sky, you will see that stars also possess subtle differences in color. These differences, particularly when analyzed using astronomical equipment, provide astronomers with enough information to infer a star's temperature, chemical composition, motion, size, age, and distance. In this module and its accompanying two tutorials and two activities, we'll explore how astronomers have uncovered the very nature of the stars.

MODULE 20. The Birth of Stars

What can we learn from multi-wavelength observations?

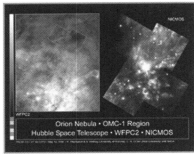

(Click image to enlarge.)

Consider the Hubble Space Telescope image of the region near Orion. The image on the left is a visible light image and the one on the right is an enlarged portion in infrared light.

Describe the main differences in observing Orion at these two wavelengths.

Observation: The Birth of Stars, page 1 of 1

If stars exist in a variety of brightnesses, sizes, colors, temperatures, and compositions, you might wonder the extent to which stars vary on the inside. Exploring fundamental questions about the nature of stars eventually will lead to the question of where and when stars originated. Although it might seem that the stars have been eternal and unchanging in the sky since the dawn of history, we now understand that stars have finite life spans—they form, they breathe, and, eventually, they die. To learn more about the origin of the stars, astronomers have to peer deep into the dust and gas clouds orbiting in a disk around our galaxy and use the principles of physics to infer how stars slowly change over time.

MODULE 21. Stellar Evolution: After the Main Sequence

Stars That Vary in Brightness

When stars become post-main sequence stars, they develop a fascinating characteristic—they vary in brightness by growing and shrinking. Such older stars that change in brightness regularly are called pulsating variable stars. How long it takes to vary in brightness is directly related to how large the star is. Large stars take a very long time to slowly grow and shrink simply due to their size. Mira variables or long-period variables, are large and cool red giant stars that take 100 days or more to go from their brightest to dimmest and back to brightest again. At the other end of the spectrum, variable stars that change from bright to dim to bright quickly are called RR Lyrae variables. These tiny stars have fluctuation periods of less than 24 hours.

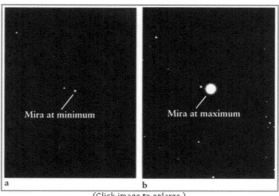

(Click image to enlarge.)

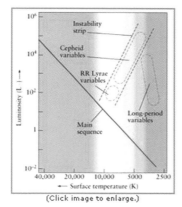

(Click image to enlarge.)

Probably the most well understood of the variable stars are the Cepheid variables (more info). These medium-sized stars take between 1 and 100 days to go from their brightest to dimmest to brightest. They are found on the H-R diagram in a region known as the instability strip. During the portion of a star's life when it remains within this region of the H-R diagram, the star will be unstable and fluctuate in size and brightness.

Pulsating variable stars are found in the upper right of the H-R diagram. Long-period variables like Mira are cool red giant stars that pulsate slowly, changing their brightness in a semiregular fashion over months or years. Cepheid variables and RR Lyrae variables are located in the instability strip, which lies between the main sequence and the red-giant region. A star passing through this strip along its evolutionary track becomes unstable and pulsates.

Cepheid variables play an important part in determining the distance to stars. Because their period of luminosity fluctuation is directly tied to their size (big stars take longer to fluctuate), the period is a direct indication of actual luminosity. This idea is called the period-luminosity relation (more info) and it is infinitely useful to the astronomer for the following simple reason. If you know the actual luminosity of a star and you go outside and measure its apparent brightness, the combination of these two pieces of information yields the actual distance to the star. For example, if you see a flashlight across the yard and decide that it looks like a 45-watt light bulb but you know that it is really a 90-watt light bulb, you can determine exactly how far away it is located. The situation is the same for determining distance in astronomy.

Tutorial: Stellar Evolution: After the Main Sequence, page 6 of 10

The twinkling main-sequence stars visible in the night sky shine by converting hydrogen into helium through thermonuclear reactions. Although stars can use this energy source for a very long time, it cannot last indefinitely. As stars approach the limit of their usable fuel resources, what happens to stars varies depending on how massive the star is to begin with and whether or not a nearby companion star exists. When our Sun reaches the end of its life, it will swell up to be more than 100 times larger and 2000 times more luminous than it is today. Some 5 billion years from now, this evolutionary phase of our star will signal the end of life as we know it on Earth. We know this swelling is what happens to stars because we can observe similar processes happening in old stars throughout our galaxy.

MODULE 22. Stellar Evolution: The Deaths of Stars

Planetary Nebulae

Low-mass stars gently eject their outer layers creating spectacular planetary nebulae.

The expanding asymptotic giant branch (AGB) star goes through a series of bursts in luminosity (more info), and in each burst it ejects a shell of material into outer space. In other words, the outer layers become detached from the core and expand into the emptiness of space. The result is an expanding ring of material surrounding the hot and isolated, but tiny and dim, core of the original star. This ring is called a planetary nebula (more info), not because it has anything to do with planets but because early observers in the 1800s with small telescopes mistook these objects for planets. These outer layers create some of the most beautiful telescope targets that can resemble colorful smoke rings. The gas shines because it is ionized by the ultraviolet radiation emanating from the exposed core. Unfortunately, these objects do not last forever. They dissipate and become an invisible part of the surrounding interstellar medium in less than 50,000 years.

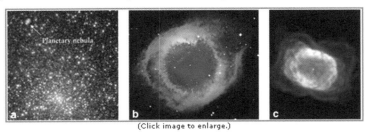

(Click image to enlarge.)

(a) The pinkish blob is a planetary nebula surrounding one of the 30,000 old stars in the globular cluster M15. The cluster is about 10,000 pc (33,000 ly) from Earth in the constellation Pegasus. (b) The closest of all known planetary nebulae is the Helix Nebula, which lies 140 pc (450 ly) from Earth in the constellation Aquarius and has an angular diameter of about 1/2° (the same as the full Moon). The "ring" is actually a spherical shell; it appears as a ring because we see a substantial thickness of the shell only when we look near its rim. (c) This infrared image of the planetary nebula NGC 7027 shows wispy shells of cooler gas surrounding a more luminous shell of hot gas, suggesting a complex evolutionary history. NGC 7027 is about 900 pc (3000 ly) from Earth in the constellation Cygnus and is roughly 14,000 AU across. (a: NASA/The Hubble Heritage Team, STScI/AURA; b: Anglo-Australian Observatory; c: William B. Latter, SIRTF Science Center/Caltech; and NASA)

Tutorial: The Deaths of Low-Mass Stars, page 4 of 8

One of the most exciting topics in astronomy revolves around the exotic deaths of stars. When stars deplete their useable fuel reserves in the core, significant changes happen in the star. Low-mass stars swell up to enormous sizes, eject their outer layers, and spend the rest of eternity as a slowly cooling sphere of ash. High-mass stars, on the other hand, go through catastrophic changes, sometimes resulting in explosions that can outshine an entire galaxy. The spectacular debris that remains from such explosions scatters the large atoms that make for stupendous sights through telescopes and can eventually combine into stars and planetary systems such as our own.

MODULE 23. Neutron Stars

Discovery of Pulsars

The first observations of neutron stars happened quite unexpectedly. A group of scientists at Cambridge University was building radio telescope antennae to scrutinize twinkling radio emissions from stars in the sky. While carefully looking at how the signals changed over time, graduate student Jocelyn Bell noticed that the pulses in one particular location in the sky fluctuated regularly. In fact, the pulses arrived one after the other exactly 1.3373011 seconds apart. These pulses were much more rapid and more regular than others known at the time. These pulsating radio signals were considered too odd to be of natural origin, and the Cambridge team at one time even considered them to have an origin in some extraterrestrial civilization. Eventually, more of the sources were discovered, and the team bestowed the name pulsar on them.

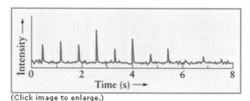
(Click image to enlarge.)

This chart recording shows the intensity of radio emission from one of the first pulsars to be discovered, PSR 0329+54. (The designation means "pulsar at a right ascension of 03h 29m and a declination of +54°"). Note that some pulses are weak and others are strong. Nevertheless, the spacing between pulses is exactly 0.714 seconds. (Adapted from R. N. Manchester and J. H. Taylor)

The radio signals detected by radio telescopes can be converted into sound and played by your computer through speakers. Access the converted signals from several famous pulsars by hyperlinking to:
http://www.jb.man.ac.uk/~pulsar/Education/Sounds/sounds.html.

Tutorial: Neutron Stars, page 4 of 8

When a massive, dying star becomes a brilliant supernova, it can emit as much light as an entire galaxy all on its own. But what happens after that? What is left after a supernova event? When a supernova occurs, most often the result is that the stellar core is crushed to nearly an unimaginable density. This object shrinks to a size much smaller than a planet, a diameter more like the size of a large city. It shines with very little light, making it quite difficult to see and even harder to study. But these objects have another curious characteristic—they spin. Some of these odd objects even spin hundreds of times each second.

MODULE 24. Black Holes

Voyage into a Black Hole

Inside a black hole, gravity distorts the structure of spacetime so severely that the directions of space and time become interchanged. At a black hole's singularity, the strength of gravity is infinite, and space and time are no longer separate entities. As a result, an imaginary spacecraft moving toward a black hole would <u>accelerate</u> as it moved nearer and nearer. Because of the intense gravitational field, the general theory of relativity becomes more apparent, and time on an imaginary spacecraft nearing a black hole would slow down. As the ship traveled closer and closer in a stronger and stronger gravitational field, time would move so slowly on the ship that the ship would never disappear into the black hole. In other words, the journey to a black hole takes forever.

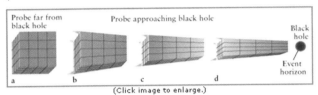

(Click image to enlarge.)

(a) A cube-shaped probe is dropped from a distance of 1000 Schwarzschild radii from a 5-$M_\odot$ black hole. (b), (c), (d) As the probe approaches the event horizon, it is distorted into a long, thin shape by the black hole's extreme gravity. A distant observer sees the probe change color as photons from the probe undergo a strong gravitational redshift.

Tutorial: Black Holes, page 8 of 10

The universe is filled with bizarre objects that result from the death throes of stars. These baffling objects include white dwarfs, planetary nebulae, and neutron stars. Without question, the strangest of these are known as black holes. Black holes are the rare result of a supernova event from the largest and most massive of stars. The only way that we can explain them manipulates both space and time in ways predicted by Albert Einstein in theories that call into question our normal ideas of what space and time are really like. The unexpected result is an object that behaves in ways that often surprise even the most experienced of scientists.

MODULE 25. Our Galaxy

Probing the Galactic Center

The clumped gas and dust that compose the visible structure of our Galaxy hamper observations that would reveal many of its secrets. For example, it is nearly impossible to determine what exists on the opposite side of our Galaxy. Further, the dust and gas make it difficult to determine what lies at the center of the Milky Way. Fortunately, most of the infrared wavelengths and radio waves emanating from the center of the galaxy do pass through the dust and gas and reach Earth. Observations at these wavelengths demonstrate that the center of the Milky Way (more info) is a very crowded place.

The center of our Galaxy is located in the direction of the constellation Sagittarius (more info). This location is an intensely strong emitter of infrared and radio waves yet takes up only a very small volume in space. It has a mass of more than 3 million solar masses but is perhaps about the size of our solar system. The only object astronomers can imagine that could harness and control this much energy is a giant black hole, named Sagittarius A* (said "Sagittarius A star"). Its gravity is strong enough to make stars move near the galactic center (more info) at incredibly high speeds exceeding 1500 km/s. Sagittarius A * is most certainly an extremely massive black hole.

(Click image to enlarge.)

(a) and (b) show the center of our Galaxy at radio wavelengths. The strength of radio emission is shown by the colors of the rainbow: Red is brightest and violet is dimmest. Black shows regions that emit negligible amounts of radio waves. (a) This view covers an area of the sky with about the same angular size as the full Moon, corresponding to a diameter of about 80 pc (250 ly). The galactic nucleus is within the rectangle in the red-colored region of strong emission. (b) This high-resolution radio view shows an area about 10 pc (30 ly) across. The material that makes up the bright pinwheel is orbiting around Sagittarius A*. (c) The cross at the center of this infrared image marks the location of the black hole at the Galaxy's center. (a, b: VLA, NRAO; c: K. M. Menten and M. J. Reid [Harvard-Smithsonian Center for Astrophysics], A. Eckart and R. Genzel [Max-Planck-Institut für extraterrestrische Physik])

Many of the best summer nights are those clear nights when the stars overhead seem to be countless. On those nights, one can sometimes see a wispy band of light stretching from the northern horizon to the southern horizon. Often mistaken for clouds, this band is more formally known as the Milky Way. It is the region of the sky where most of the visible stars are located. Although our eyes cannot resolve most of these dim and distant stars, we can see the collective light from them as a streak of dim light across the sky. A telescope, particularly one that focuses on nonvisible wavelengths, is able to see many of these stars and, with some scientific inference, can reveal the structure of our galaxy.

MODULE 26. Galaxies

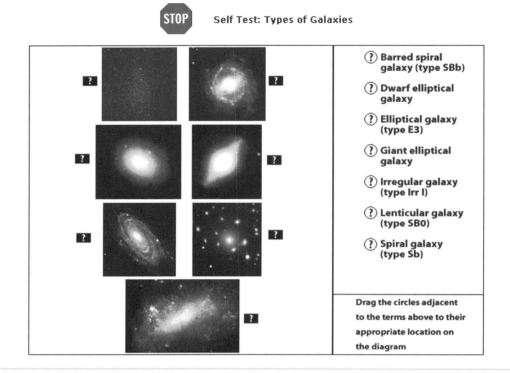

Some of the most dazzling pictures in astronomy are of isolated islands of millions, even billions, of stars. These islands are known as galaxies and are found in an amazing array of shapes and sizes. Some are highly structured with distinct and curving arms while others have no distinguishable shape at all. Some are round while others are long and slender. But whatever style is the most visually appealing to you, one thing is certain—there are plenty of them. There are more than 100 billion galaxies in our universe and possibly many more. They are found in every direction, in every size, and at many different stages of maturity.

MODULE 27. Quasars, Active Galaxies, and Gamma-Ray Bursters

Gamma-Ray Bursters

GRBs were first discovered by accident. In the 1960s, the United States positioned a series of orbiting satellites called Vela to look for the high-energy releases that happen when nuclear weapons are detonated. Since the discovery of GRBs in the heavens, satellites such as the Italian-Dutch BeppoSAX and NASA's SWIFT mission have been designed to patiently wait for a GRB to occur. When a GRB does happen, these satellites quickly pinpoint the originating position and help other telescopes to look in the same direction. In this way, we now know that GRBs originate billions of light-years away and for that reason must be caused by tremendously energetic events. One possible cause is that GRBs are created when two neutron stars collide, an event affectionately dubbed a *hypernova*. Another possibility is that GRBs are created when plasma from a small portion of a supernova event is superheated. The astronomical jury is still out about the exact process that produces the observed GRBs

(Click image to enlarge.)

The Hubble Space Telescope recorded this visible-light image on February 8, 1999, 16 days after a gamma-ray burster was observed at this location. The image shows a faint galaxy that is presumably the home of whatever object produced the gamma-ray burst. The galaxy has a very blue color, indicating the presence of many recently formed stars. The gamma-ray burst may have been produced when one of the most massive of these stars became a supernova. (Andrew Fruchter, STScI; and NASA)

Tutorial: Quasars, Active Galaxies, and Gamma-Ray Bursters, page 7 of 8

The stars and galaxies that make up our night sky typically emit most of their energy in the ultraviolet, visible, and infrared wavelength ranges in a manner that is directly related to their temperatures. There are, however, a few objects in the cosmos that do not obey this generalization. These are objects with odd names such as quasars and blazars that emit enormous amounts of energy at a variety of wavelengths. Many of these objects are astoundingly luminous and yet sometimes exist as a highly compact source. Exactly what these objects may be presents many of the most perplexing mysteries to today's astronomers.

MODULE 28. Cosmology: The Creation and Fate of the Universe

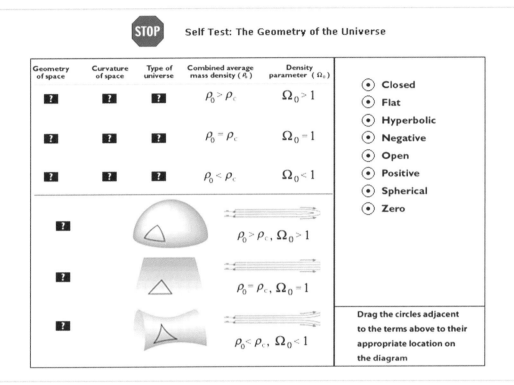

Tutorial: Cosmology, page 6 of 9

Most cultures pose questions about where the universe came from and where is it going: How did the cosmos that we see begin? What will happen to them? The enterprise of science is only different in that the questions posed by scientists are addressed using the logical and systematic methods of science. Cosmology is the name scientists use to describe the study of the history of the evolution of the universe as a whole. *Universe* means absolutely everything that exists in both space and time. To understand the rise (and possible decline) of our universe, cosmologists use the most modern technology to study changes in the most distant galaxies as well as the behavior of the smallest subatomic particles. Contemporary cosmology is an exciting endeavor that often presents many more questions than answers.

MODULE 29. Exploring the Early Universe

Cosmic Strings

Spontaneous symmetry breaking plays an important role in thinking about what things were like in the first few moments of the universe. Approximately 10^{-43} seconds after the Big Bang, the temperature of the universe dropped to a point that the gravitational force became separated from the electromagnetic, strong, and weak forces. Another way to say this is that the gravitational force was "frozen out." Similarly, when the temperature cooled to 10^{27} K, the strong force "froze out." When this occurred, some places in the universe were at a substantially higher energy state than other places, and it is conceivable that some places actually were able to avoid symmetry breaking and retain the combination of the forces. These places have become known as cosmic strings. Cosmic strings would be incredibly massive but invisible, and they may account for some of the dark matter in the universe.

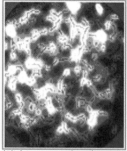

(Click image to enlarge.)

This map shows the locations of nearly half a million galaxies spanning 100° of the sky. The north pole of our Galaxy (in the constellation Coma Berenices) is at the center. Each small square covers 10 x 10 arcmin. White indicates a high density of galaxies, and darker reddish regions indicate few galaxies. Green squares emphasize filaments along which galaxies are concentrated, and the single red squares (near the centers of the white regions) indicate the points where galaxy counts reach a maximum. (Courtesy of E. L. Turner, J. E. Moody, and J. R. Gott)

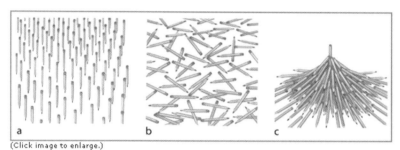

(Click image to enlarge.)

(a) The symmetry that existed in the early universe can be likened to orderly rows of pencils balanced on their points. (b) The state of broken symmetry that exists today is analogous to all the pencils having fallen over. (c) A situation could arise, however, in which the falling pencils might topple toward each other so that one pencil is supported upright. Some physicists theorize that comparable structures, called cosmic strings, might be left over from the symmetry breaking that occurred during the earliest moments of the universe.

Tutorial: Exploring the Early Universe, page 7 of 9

Most people realize that the endeavor of cosmology relies on the study of giant galaxies at astronomical distances, but rarely are people aware of how much cosmology depends on the study of subatomic particles. The behavior of strange subatomic particles, such as electrons and positrons, has had an enormous impact on the structure of the universe right from the first instant of time. Advances in our knowledge of subatomic particles have primarily come from giant particle accelerators that make atoms collide together. Yet even our most powerful accelerators cannot match the energies that were present in the very early universe.

MODULE 30. The Search for Extraterrestrial Life

Planets Orbiting Other Stars

Access the most recent discoveries of extrasolar planets at
http://exoplanets.org/almanacframe.html, and use the data there to answer the following questions.

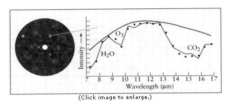

(Click image to enlarge.)

The image above is a simulation of what the Terrestrial Planet Finder infrared telescope might see when it is deployed around 2011. The white dot at the center is a nearby Sunlike star, and the smaller dots around it are planets orbiting the star. On the right is the simulated infrared spectrum of one of the planets, showing broad absorption lines of water vapor (H_2O), ozone (O_3), and carbon dioxide (CO_2). While all these molecules can be created by nonbiological processes, the presence of life will change the relative amounts of each molecule in the planet's atmosphere. Thus, the infrared spectrum of such planets will make it possible to identify worlds on which life may have evolved. (Jet Propulsion Laboratory)

1. For the last three most recent planets discovered orbiting stars beyond our solar system:

   a. Which ones have masses greater than that of our planet Jupiter (M_{jup})?

   b. Which ones orbit their star closer than Earth orbits the Sun (1 AU)?

   c. How circular (eccentric) are their orbits compared to Earth's orbit (e = 0.017)?

2. What generalizations can you make about all the planets discovered so far in terms of mass, distance to the Sun, and orbit shape?

   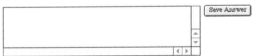

To save your answer, click the "Save Answer" button before you leave this page. When you are finished with this module, go to the **Module Conclusion** and follow the instructions for sending this information to your course instructor.

Observation: The Search for Extraterrestrial Life, page 1 of 1

One of the most exciting endeavors of science of the last ten years is the search for extraterrestrial life. Known as astrobiology, or sometimes as exobiology, this new scientific discipline promotes the interaction of astronomy, biology, chemistry, geology, mathematics, and physics. The main goal of astrobiology is to understand the origin and evolution of life throughout the universe. The science of astrobiology has been catapulted into the public spotlight in recent years following the discovery of life in places once thought completely inhospitable—at the bottom of the ocean, in boiling-hot sulfur springs, and in the frigid Antarctic desert. At the same time, we have discovered more planets orbiting other stars than exist in our own solar system. These recent discoveries, in concert with radio searches for intelligent civilizations such as those led by SETI (Search for Extraterrestrial Intelligence), make astrobiology one of the most exciting fields in science today.

Notes